Velocity and Vision

How Extreme Sports Inspire Creativity and Define Generations

Winona Otero

Contents

1 **Introduction: The Rush of Innovation** **5**
1.1 Defining Extreme Sports 7
1.2 Creativity's Edge 9

2 **Chapter 2: Adrenaline & Art** **13**
2.1 Risk and Reward 16
2.2 Pushing Boundaries 18
2.3 Artistic Expression 20

3 **Chapter 3: The X-Games Generation** **23**
3.1 Media's Influence 25
3.2 Youth Culture 27
3.3 Impact on Society 29
3.4 Technological Advancements 31
3.5 Future Trends 33

4 **Chapter 4: Beyond the Physical** **37**
4.1 Mental Fortitude 39
4.2 Overcoming Fear 42

5 **Chapter 5: Innovation in Design** **45**
5.1 Equipment Evolution 47
5.2 Safety and Technology 50
5.3 Sustainable Practices 52

6 **Chapter 6: Global Impact** **55**
6.1 International Competitions 57
6.2 Cultural Exchange 60
6.3 Economic Influence 62
6.4 Environmental Concerns 64
6.5 Community Building 67

7 **Chapter 7: The Future of Extreme** **71**
7.1 Emerging Sports 74
7.2 Technological Frontiers 76

1 Introduction: The Rush of Innovation

Extreme sports: They're not just about athleticism; they're a crucible forging innovation. Think about it – the constant push for higher jumps, faster speeds, and more daring maneuvers necessitates constant refinement, adaptation, and the creation of entirely new approaches. This relentless pursuit of the edge breeds a culture of invention, one where limitations are seen not as barriers, but as challenges to be overcome creatively. This innovative spirit isn't confined to the athletes themselves; it ripples outward, influencing design, technology, and even our understanding of human potential.

The very definition of "extreme" is fluid, constantly shifting as athletes push boundaries and redefine what's possible. This inherent instability demands flexibility and adaptability, traits crucial for innovative thinking in any field. The willingness to embrace risk, to accept failure as a learning opportunity, is a cornerstone of both extreme sports and groundbreaking innovation. It's a mindset that values experimentation over complacency, a leap of faith over cautious calculation. This iterative process of trial and error, of pushing limits and refining techniques, is what drives progress.

Consider the evolution of snowboarding. From its humble beginnings as a quirky alternative to skiing, it's exploded into a multifaceted sport with diverse disciplines, each demanding specialized equipment and techniques. This evolution wasn't a linear progression; it was a chaotic dance of experimentation, fueled by the desire to conquer new terrain and express individuality. Similarly, the development of climbing gear, from rudimentary ropes to advanced belay systems and specialized footwear, showcases a relentless drive for improved safety and performance.

The inherent risk in extreme sports necessitates a deep understanding of physics, engineering, and materials science. Innovations in equipment and safety technology are not merely incremental improvements; they often represent significant leaps forward in these fields. For example, advancements in helmet design, harness technology, and protective gear have dramatically reduced the risk of injury, allowing athletes to push their limits with greater confidence. This constant push for safety mirrors the innovative spirit found in other high-risk fields, like aerospace engineering or medical technology.

Moreover, the tight-knit community fostered within extreme sports contributes to a collaborative environment where ideas are shared freely, and innovation is a collective effort. Athletes, designers, and engineers work together, constantly exchanging knowledge and pushing each other to develop new techniques, equipment, and approaches. This symbiotic relationship fuels a cycle of continuous improvement, where the boundaries of what's possible are constantly being redrawn. This collaborative spirit is a powerful catalyst for innovation, demonstrating the importance of shared knowledge and open communication in driving progress.

The rapid pace of technological advancement further accelerates this cycle. High-speed cameras, advanced data analytics, and biomechanical analysis are increasingly used to study athletic performance, providing valuable insights that inform the development of new techniques and equipment. This feedback loop between technology and athletic performance represents a powerful synergy, driving continuous innovation and pushing the boundaries of human capabilities. This isn't just about enhancing performance; it's about refining the very understanding of movement, skill, and the human body's potential.

Beyond the physical, the mental demands of extreme sports cultivate a resilience and creative problem-solving capacity that translates to success in other areas of life. The ability to overcome fear, manage risk, and maintain focus under pressure are valuable assets in any challenging endeavor. This mental toughness, honed through years of rigorous training and competition, equips individuals with the inner strength needed to tackle complex problems and innovate effectively in the face of adversity. These are transferable skills, applicable not just in the athletic arena, but in business, science, and every facet of human endeavour.

Finally, the very act of pushing boundaries, of striving for something seemingly impossible, is in itself an act of creation. It's a testament to the human spirit's relentless drive to explore, discover, and innovate. The adrenaline rush, the sense of accomplishment, the feeling of pushing past perceived limitations – these are the forces that fuel the creative spark, driving athletes and innovators alike to continuously surpass expectations. Extreme sports, therefore, represent not just a pursuit of physical excellence but a vibrant expression of human ingenuity and a powerful catalyst for innovation across diverse fields.

1.1 Defining Extreme Sports

Defining extreme sports requires moving beyond simple definitions. It's not merely about high speeds or dangerous stunts; it's a nuanced concept encompassing a specific blend of risk, skill, and a unique relationship with the environment. Traditional sports often involve structured rules, predictable environments, and a focus on competition within established parameters. Extreme sports, conversely, frequently thrive in less controlled settings, emphasizing improvisation, adaptability, and personal mastery against the odds, often in the face of nature's unpredictable forces. Think of a surfer battling monstrous waves, a climber scaling a sheer cliff face, or a base jumper leaping from a towering structure. These are not games; they are confrontations with nature's raw power, tests of courage, and explorations of human potential. The very essence of the sport lies in the inherent danger, the palpable awareness of potential injury or death, which heightens the sense of accomplishment and the intensity of the experience. It is this interplay of controlled risk-taking and mastery of skills that forms the core of what defines extreme sports.

The element of risk is not simply a side effect; it's integral to the experience. The adrenaline surge, the intense focus required to overcome fear, and the sense of accomplishment following a successful attempt are defining characteristics. This risk-reward dynamic attracts participants seeking something beyond traditional competition. It's about pushing personal boundaries, achieving feats previously thought impossible, and testing one's physical and mental limits in a way that traditional sports simply cannot match. This desire to confront personal limitations and redefine perceived boundaries is precisely what elevates extreme sports beyond mere recreation, transforming them into compelling displays of human tenacity and skill. The line between success and failure is often razor-thin, adding another layer to the drama and intrigue of these disciplines. This makes the athletes not only incredibly skilled but also profoundly courageous.

Furthermore, the definition expands to incorporate the location and context. Unlike many organized sports, many extreme sports are defined by the environment itself. Mountain biking isn't just about riding a bike; it's about navigating treacherous terrain, rock faces, and steep descents. Similarly, skateboarding pushes the boundaries of urban landscapes, transforming ordinary concrete into a playground for incredible feats of balance and agility. This close relationship with the natural world, or with human-made environments, is crucial to understanding extreme sports; the environment itself is not merely a setting but an active participant in the event, shaping the experience and adding layers of complexity and challenge. This interconnectedness between the athlete, their skills, and the environment sets them apart.

However, it's crucial to acknowledge the subjective nature of the term "extreme." What constitutes "extreme" for one person may be commonplace for another. This fluidity underscores the evolving nature of extreme sports. As skills improve and athletes push boundaries, previously considered "extreme" feats become more commonplace. New disciplines emerge, pushing the definition even further. This inherent dynamism means that the definition remains fluid, perpetually reshaped by technological advancements, human ingenuity, and the unwavering human desire to test limits. The ever-shifting boundaries of what's possible are perhaps the most extreme aspect of extreme sports. Consequently, defining the term requires acknowledging this inherent evolution, understanding it not as a fixed entity but as a continuously unfolding concept reflecting human ambition and the ceaseless quest for new challenges. That, in itself, is what makes it truly extreme.

1.2 Creativity's Edge

The inherent link between extreme sports and creativity isn't merely anecdotal; it's deeply rooted in the very nature of the activities themselves. Consider the constant problem-solving demanded by a skateboarder navigating a complex urban landscape, or a rock climber meticulously planning each handhold and foothold. These aren't just physical feats; they're creative acts, demanding improvisation, adaptation, and a constant reevaluation of the environment and one's own capabilities. The athlete isn't simply reacting; they're actively creating their own performance, their own narrative within the constraints of the sport.

This creative process is further amplified by the intense pressure and risk involved. Under duress, our minds often operate differently, accessing unconventional solutions and strategies that might remain dormant under less stressful conditions. The adrenaline surge associated with these sports isn't simply a physiological response; it fuels a heightened state of awareness and focus, sharpening instincts and prompting innovative thinking. This isn't to say that risk-taking alone guarantees creativity, but the intense focus and quick decision-making processes necessitate a high degree of creative problem-solving.

Furthermore, the visual aesthetics inherent in many extreme sports should not be overlooked. The fluidity of a snowboarder carving down a mountain, the precise movements of a surfer riding a wave, the graceful power of a BMX rider executing a complex trick – these are artistic expressions as much as they are athletic achievements. The athletes are not just performers, they're choreographers, constantly seeking to refine their movements, to find new ways to express themselves within the boundaries of their chosen discipline. This artistic impulse underscores the inextricable link between physical prowess and creative expression.

This creative drive extends beyond the individual athlete to influence the broader community. The continuous pursuit of innovation in equipment design, for example, demands a relentless creative process. From the evolution of skateboards to the development of safer climbing gear, the constant drive to improve performance and reduce risk necessitates creative solutions in material science, engineering, and technology. The design process itself is one of continuous experimentation, testing, and refinement, pushing the boundaries of what's possible.

The influence of extreme sports extends beyond material innovation. Consider the impact on media and culture. The visual spectacle of extreme sports lends itself readily to captivating media coverage, generating widespread interest and influencing popular culture in unexpected ways. This influence is evident in music, fashion, and even language, with the adoption of extreme sports terminology and imagery into mainstream society. This broader cultural influence highlights the powerful creative force behind the sports.

Moreover, the community built around extreme sports is itself a testament to the power of creative collaboration and innovation. Athletes, designers, filmmakers, and other individuals constantly interact and exchange ideas, fueling a dynamic creative ecosystem. This collaborative environment fosters the sharing of knowledge, the development of new techniques, and the generation of innovative ideas, highlighting the synergistic nature of the creative process within the context of these high-risk sports. The constant exchange of information and experiences continually pushes the boundaries of what is considered possible, leading to breakthroughs both on and off the field.

Ultimately, the creativity fostered within extreme sports provides a unique lens through which we can understand the human capacity for innovation. It demonstrates that creativity isn't solely confined to traditional art forms or intellectual pursuits; it is a fundamental aspect of the human experience, capable of manifesting itself in diverse and unexpected ways. The daring spirit, relentless pursuit of self-improvement, and innovative problem-solving exhibited by extreme sports athletes serve as a powerful inspiration for us all, reminding us of the boundless potential for creativity within ourselves. The next generation of innovators will likely be significantly influenced by the spirit and ideals embodied in the world of extreme sports; a world where creativity isn't merely a pursuit, but a fundamental necessity. The very act of participating demands a continual creative response. The pursuit of extreme sports inherently fosters and demands an extreme level of creativity.

2 Chapter 2: Adrenaline & Art

The inherent risk in extreme sports isn't simply a reckless disregard for safety; it's a calculated gamble, a dance with danger that yields extraordinary rewards. This calculated risk fosters a unique kind of creativity, forcing athletes to solve complex problems under immense pressure, adapting strategies on the fly, and refining their skills to a level of precision few other disciplines demand. The payoff isn't just physical mastery; it's a profound understanding of one's own capabilities and limitations, a humbling experience that fuels personal growth. This constant push for improvement isn't solely about winning medals, but about the exhilaration of pushing boundaries, the satisfaction of conquering fear, and the ongoing pursuit of self-discovery.

Consider the graceful arc of a snowboarder carving through fresh powder, the fluid movements of a surfer navigating a monstrous wave, or the breathtaking precision of a rock climber ascending a sheer cliff face. These aren't merely athletic feats; they are acts of artistic expression, each movement a brushstroke on the canvas of the natural world. The athletes become choreographers of their own bodies, translating risk and adrenaline into a performance of controlled chaos, a symphony of skill and courage. Their bodies become instruments, and the environment, their stage. This isn't about conquering nature, but harmonizing with it, finding a rhythm and a flow that transcends mere physical prowess.

The adrenaline surge itself plays a pivotal role in this creative process. The heightened state of awareness, the razor-sharp focus, and the intense emotional experience create an environment conducive to innovative thinking. It's in these moments of heightened sensory input and physiological response that groundbreaking solutions often emerge, intuitive maneuvers appearing as if by magic. This heightened state isn't about reckless abandon; it's about harnessing the power of the body's response to pressure, transforming it from a potential impediment to a driving force. This is where the true artistry resides, not only in the performance itself, but in the mental and emotional resilience required to manage it.

This creative process isn't solely confined to the athletes themselves; it extends to the designers and engineers who craft the equipment that makes these feats possible. Every advancement in materials science, every innovation in design, represents a response to the athlete's relentless push for excellence. The development of lighter, stronger, more durable equipment, driven by the demands of extreme athletes, has had a ripple effect, inspiring innovation in various other fields. The need for increased safety features, for example, has pushed the boundaries of protective gear technology, leading to advancements that benefit not only extreme sports participants, but also other high-risk professions. This symbiotic relationship between athlete, technology, and art continues to redefine what's possible.

Moreover, the visual impact of these extreme sports cannot be ignored. The stunning landscapes, the breathtaking stunts, the raw emotion captured in photographs and videos – these all contribute to a powerful visual language that transcends cultural boundaries. The imagery speaks to our primal instincts, our fascination with risk, our admiration for skill and resilience. This visceral experience fosters a connection between the athlete and the spectator, a shared understanding of the thrill and beauty that is found in pushing one's limits. This shared experience has shaped aesthetic trends, influenced artistic expressions, and served as inspiration for countless creative works. The adrenaline, the risk, the sheer beauty of the sports – these elements are all inextricably linked, creating a visual narrative that resonates deeply.

The aesthetic quality of extreme sports also extends beyond the visuals. The sounds of wind, waves, and rushing water combine with the rhythmic thuds and whooshes of equipment, creating a soundscape as dynamic and intense as the visuals. The combination of the visual and auditory elements elevates the experience, not just for the athlete, but also for those watching or participating vicariously. This shared aesthetic appreciation helps to foster a community of like-minded individuals, bound together by a common passion for risk, adrenaline, and artistic expression in extreme sports. They are united by the exhilaration of the experience, irrespective of their personal level of involvement.

Finally, the inherent drama and storytelling potential of extreme sports translate seamlessly into artistic expression. Documentaries, films, and literary works explore the intense physical and mental challenges faced by athletes, the sacrifices they make, and the triumphs they achieve. These narratives often highlight the human spirit's capacity for resilience, courage, and determination. They encapsulate the struggle against fear, the pursuit of excellence, and the unwavering commitment to a dream. The artistic interpretations of these stories, whether through film, literature, or visual art, broaden the reach and impact of extreme sports, enriching our understanding of human potential and inspiring others to push their own boundaries.

2.1 Risk and Reward

Calculating the precise risk-reward ratio in extreme sports is inherently difficult; it's a subjective equation constantly recalibrated by individual athletes based on experience, intuition, and the ever-shifting landscape of their chosen discipline. For a beginner, however, understanding this fundamental principle is paramount. Ignoring risk entirely is reckless, but overemphasizing it paralyzes progress. The sweet spot lies in informed risk assessment – a careful weighing of potential gains against potential losses, a continuous process of learning and adaptation.

This necessitates a thorough understanding of your own physical and mental capabilities. Beginners frequently overestimate their abilities, fueled by adrenaline and a thirst for excitement. This leads to ill-advised attempts beyond their skill level resulting in injury, setbacks, or worse. Honest self-assessment is crucial, demanding humility and a willingness to acknowledge limitations. Start slowly, focus on mastering the basics, and gradually increase the difficulty only as your proficiency grows. Don't let impatience override common sense.

Mentorship plays a crucial role in navigating the risk-reward balance. Experienced athletes can provide invaluable guidance, helping beginners identify potential hazards, refine techniques, and develop sound judgment. Learning from those who have already traversed the learning curve accelerates progress while minimizing risk. Seek out experienced mentors, heed their advice, and embrace the opportunity to learn from their successes and, equally importantly, their mistakes. Their insights can be your most valuable asset.

Equipment is another critical component of risk management. Investing in high-quality, well-maintained gear significantly reduces the probability of equipment failure, a major source of accidents in many extreme sports. Properly functioning equipment enhances control, allowing athletes to execute maneuvers more precisely and respond effectively to unexpected situations. Cutting corners on equipment is a false economy, potentially leading to far greater costs down the line. Remember that your safety is paramount. Treat safety precautions as crucial as training.

The environment plays a considerable role in determining the risk-reward profile of a particular activity. Weather conditions, terrain, and other environmental factors can dramatically alter the level of risk involved. Understanding weather patterns, assessing terrain challenges, and being acutely aware of environmental hazards are skills that all extreme sport enthusiasts must acquire. Adaptability is key; be prepared to postpone or even cancel activities when conditions become too hazardous. Respect nature's power.

Finally, mental preparation is just as critical as physical training. Extreme sports demand mental fortitude, discipline, and the ability to manage fear. Cultivating a strong mental game involves developing strategies for managing stress, staying focused under pressure, and maintaining composure during challenging situations. Mindfulness techniques, visualization exercises, and psychological training can help enhance mental resilience and improve performance in high-stakes scenarios. The mental game is a vital component of success. Mastering it significantly reduces your risk. Never underestimate the importance of your mind.

The thrill of extreme sports lies in pushing boundaries, but a sensible approach to risk management ensures that the pursuit of reward doesn't become a gamble with irreversible consequences. By carefully weighing risks, respecting limits, learning from experience, and continuously assessing your capabilities, you can maximize your chances of success and enjoy the exhilaration of these exciting pursuits to the fullest. Embrace the challenge, but do it wisely. The reward is far sweeter when it's earned through careful consideration and calculated risk-taking. Never forget that safety and calculated risks go hand in hand.

2.2 Pushing Boundaries

Pushing boundaries isn't simply about physical feats; it's a profound mental and emotional journey. It requires a deep understanding of your own limitations, a willingness to confront fear, and an unwavering belief in your potential. This isn't about reckless abandon, but calculated risk-taking, a fine balance between pushing the envelope and ensuring your safety. Success hinges on meticulous preparation, both physically and mentally. Visualizing success, practicing relentlessly, and developing coping mechanisms for setbacks are crucial elements of this process. Athletes often employ mindfulness techniques, meditation, or visualization exercises to prepare themselves mentally for the extreme challenges they face.

The constant push to excel demands resilience. Failures are inevitable; they are not endpoints, but stepping stones. Learning from mistakes, adapting strategies, and maintaining a positive mindset, even in the face of adversity, are critical for sustained progress. The ability to analyze performance, identify weaknesses, and continuously improve are vital skills for any extreme sports athlete. It's about recognizing that growth often occurs outside of the comfort zone.

Consider the evolution of snowboarding. Early boards were cumbersome and lacked the maneuverability of modern designs. Innovation in materials and design, driven by the constant desire to push boundaries, has led to lighter, more responsive boards, enabling riders to perform increasingly complex maneuvers. This constant refinement, a testament to the relentless pursuit of progress, is a hallmark of extreme sports. Similarly, advancements in safety equipment, from helmets and protective padding to specialized clothing, have mitigated risks, allowing athletes to explore their limits with increased confidence.

The psychological aspects are equally critical. Overcoming fear isn't about its eradication; it's about managing it. Athletes learn to identify their fear triggers and develop strategies to counteract them. This might involve deep breathing exercises, positive self-talk, or simply focusing on the task at hand. The ability to control fear, rather than letting it control you, is a defining characteristic of successful extreme athletes. These individuals possess an incredible capacity for self-belief, a confidence that allows them to tackle seemingly impossible challenges.

Pushing boundaries also fosters creativity. The need to overcome obstacles, to find new and innovative ways to perform, ignites the creative process. This is where the artistry of extreme sports becomes apparent. The graceful movements of a freestyle skier, the precision of a rock climber, or the daring maneuvers of a BMX rider are not simply acts of athleticism; they're expressions of artistic skill and creativity. Athletes often design their own maneuvers, pushing the boundaries of what's considered possible, and in turn, inspiring others to do the same.

The inherent risk in extreme sports has fostered a community centered around shared experiences, mutual support, and a deep understanding of the challenges involved. This creates a strong bond between athletes, coaches, and supporters, fostering a sense of camaraderie and mutual respect. This community becomes a crucial support system, offering encouragement, guidance, and a shared understanding of the sacrifices involved in pushing boundaries.

Ultimately, the drive to push boundaries in extreme sports is more than a pursuit of adrenaline or victory. It's a testament to human resilience, creativity, and the unwavering pursuit of excellence. It's a journey of self-discovery, where athletes constantly confront their limitations and emerge stronger, more confident, and more capable. It's a process of continuous learning, adaptation, and growth, all driven by an insatiable desire to explore the limits of what's possible, both physically and mentally. The pursuit of this pushes the athlete to explore not only their own physical and mental capabilities but also the boundaries of their creativity and ingenuity. This continuous cycle of challenge and innovation is what makes extreme sports such a compelling and inspiring field of study. It is a powerful illustration of the human spirit's relentless drive to overcome obstacles and push forward into the unknown.

2.3 Artistic Expression

The raw energy of a skateboarding ollie, the calculated grace of a snowboarder carving through fresh powder, the breathtaking precision of a BASE jumper navigating a cliff face – these aren't just feats of athleticism; they're artistic statements. Each movement, each risk, each breathtaking moment is a carefully constructed expression of skill, creativity, and an individual vision. Extreme sports aren't solely about conquering physical challenges; they're about crafting a performance, a visual narrative etched against the backdrop of natural landscapes or purpose-built arenas.

Consider the fluidity of a surfer riding a wave, the way they seem to become one with the ocean's rhythm. This isn't just about staying upright; it's about interpreting the ocean's language, understanding its power, and expressing that understanding through a dance of skill and balance. This synergy, this symbiotic relationship between athlete and environment, elevates the performance beyond mere athletic competition; it becomes an aesthetic experience.

The visual aspects are key. Think about the vibrant colors of clothing and equipment, carefully chosen to create a dynamic and eye-catching display. The choreography of movement, whether it's the synchronized precision of a BMX team or the freestyle artistry of a slackliner, adds another layer of artistic expression. This intentional aesthetic is paramount; athletes frequently spend considerable time planning their routines, selecting locations, and coordinating their gear. Their goal is not merely to execute maneuvers, but to create an unforgettable spectacle.

Furthermore, photography and videography play crucial roles in amplifying this artistic expression. Slow-motion replays capture the elegance of a perfectly executed maneuver, revealing subtle details and underscoring the athlete's control and precision. Creative editing techniques can further enhance the impact, transforming raw footage into compelling visual narratives that resonate with audiences far beyond the immediate event. These visual narratives shape the perception of extreme sports, influencing public attitudes and inspiring new generations of athletes.

This emphasis on visual aesthetics is reflected in the design of extreme sports equipment itself. From the vibrant colors of skateboards to the aerodynamic designs of mountain bikes, the equipment itself becomes part of the artistic expression. Companies invest heavily in research and development, pushing the boundaries of materials science and engineering to create equipment that is not only functional but also visually stunning. This merging of form and function is integral to the overall artistic presentation.

Moreover, the music and soundtracks that accompany extreme sports videos often further enhance the aesthetic. Carefully chosen music can amplify the emotional impact of a particular sequence, guiding the viewer's emotions and creating a sense of drama and excitement. The synergy between movement and music creates a unified artistic statement that transcends mere documentation.

Beyond the visual and auditory aspects, there's a profound emotional dimension to artistic expression in extreme sports.

The courage to confront fear, the dedication to master difficult skills, and the unwavering pursuit of excellence all contribute to the narrative. This raw emotion, this palpable sense of risk and reward, is what makes extreme sports so compelling to both athletes and spectators.

Finally, it is important to remember that the creative process is intensely personal. Each athlete approaches the task with their own unique style, pushing creative boundaries in their individual ways. This individual approach ensures the continued evolution of extreme sports as an artistic medium, constantly generating fresh ideas and pushing the creative envelope. The constant evolution reflects the diversity and creativity within the global extreme sports community. The individual expression is the heart of the artistic endeavor in these sports.

3 Chapter 3: The X-Games Generation

ESPN's coverage transformed skateboarding from a niche counter-culture activity into a mainstream spectacle. Suddenly, teenagers idolized Tony Hawk, not just for his skills, but for his entrepreneurial spirit and media savvy. This amplified the sport's visibility, attracting a new generation of participants and shaping their aspirations. The X Games, born from this media explosion, solidified the connection between extreme sports and youth culture.

The X Games weren't just competitions; they were meticulously crafted media events, complete with elaborate sets, music, and storylines. This carefully constructed narrative of risk, rebellion, and youthful exuberance resonated powerfully with a generation coming of age in the digital age. Social media further amplified this effect, turning athletes into instant celebrities with millions of followers, each a potential sponsor or brand ambassador. The symbiotic relationship between extreme sports and media continues to evolve, shaping the industry's growth and influencing athletes' careers in profound ways.

This cultural shift significantly impacted societal attitudes towards risk and individual expression. The rebellious spirit often associated with extreme sports challenged conventional norms and encouraged a more adventurous outlook among young people. The visibility of athletes overcoming adversity—be it physical injury or personal challenges—created inspiring narratives that resonated far beyond the confines of the sporting world. This has been reflected in shifts in fashion, music, and even advertising, borrowing heavily from the aesthetics and ethos of extreme sports culture.

Technological advancements played a crucial role in the rise of the X-Games generation. High-definition cameras captured breathtaking footage, showcasing the athleticism and artistry of these sports in unprecedented detail. Slow-motion replays, dramatic editing, and innovative camera angles transformed extreme sports into visually stunning spectacles, further amplifying their appeal to a mass audience. Simultaneously, advancements in protective gear, equipment design, and training methodologies have allowed athletes to push boundaries further and with greater safety, fueling the evolution of the sports themselves.

Predicting the future of extreme sports is inherently speculative, yet some trends are clear. The continued influence of digital media will shape how future generations engage with these sports, fostering new forms of participation and creating entirely new avenues for creativity and competition. Virtual and augmented reality technologies, already showing promise, may well transform the landscape of training, competition, and spectator engagement. The ethical considerations surrounding safety, environmental impact, and the commercialization of extreme sports will require careful consideration moving forward. We can expect greater professionalization, increased sponsorship opportunities, and a continuing evolution of both the sports and the media ecosystem that surrounds them. Ultimately, the future of extreme sports is likely to be even more dynamic, technologically advanced, and globally connected than its present state. This will create an interesting scenario where not only the sports themselves evolve but also the expectations and perceptions surrounding these intense activities. The generation shaped by this evolution is poised to be as impactful as those before them, further pushing the boundaries of both athletic achievement and media influence.

3.1 Media's Influence

Media's pervasive influence on extreme sports is undeniable, shaping its evolution from niche activities to a globally recognized phenomenon. Early exposure primarily came through word-of-mouth and limited news coverage, fostering an air of mystery and rebelliousness. This underground appeal, however, rapidly changed with the advent of television and video technology. The inherent spectacle of extreme sports—the breathtaking stunts, daring maneuvers, and often-perilous environments—proved irresistibly captivating to viewers, leading to a surge in popularity.

The rise of dedicated television channels, such as ESPN's X Games, marked a pivotal moment. These broadcasts brought extreme sports directly into people's homes, transforming previously obscure athletes into household names. This newfound exposure wasn't just about showcasing athletic prowess; it carefully crafted narratives around individual athletes, emphasizing their personalities, backstories, and personal struggles. These meticulously crafted storylines, often mirroring heroic archetypes, profoundly influenced public perception, creating aspirational figures that resonated with audiences worldwide.

Consequently, the media's portrayal extended beyond the athletes themselves. Sponsors eagerly jumped on board, recognizing the marketing potential of associating their brands with the excitement and rebellious spirit of extreme sports. This symbiotic relationship propelled the growth of the industry, funding equipment development, athlete training, and the expansion of competition circuits. However, this commercialization also brought about concerns regarding the authenticity and integrity of the sport. Some argue that the focus on sponsorships and endorsements diluted the raw, adventurous spirit that initially characterized extreme sports.

The digital age has further amplified media's impact. The internet, social media platforms, and video-sharing websites offer athletes unprecedented opportunities for self-promotion and direct engagement with fans. This direct line of communication fosters a sense of intimacy and allows athletes to cultivate a personal brand, independent of traditional media gatekeepers. However, it also introduces new challenges. The pressure to maintain a constant online presence, to create engaging content, and to manage their public image constantly, can be extremely demanding on athletes' mental health.

Furthermore, the ease of content creation and distribution via the internet has led to a proliferation of amateur and user-generated content. This democratization of media allows everyday enthusiasts to share their experiences, fostering a sense of community and expanding the reach of extreme sports beyond the elite level. But it simultaneously opens up the potential for the spread of misinformation and the perpetuation of unrealistic expectations. The carefully edited highlights on social media seldom convey the dedication, training, and inherent risks involved in extreme sports.

In closing, while media has played a crucial role in popularizing and shaping extreme sports, its influence requires critical examination. The commercialization and the constant pressure for online engagement can be double-edged swords. Yet, the increased accessibility provided by digital media platforms continues to broaden the participation base and nurture the vibrant global community that surrounds this unique and dynamic athletic field. Understanding this complex relationship between extreme sports and media is essential to fostering responsible growth and appreciating its multifaceted impact.

The future of extreme sports will undoubtedly be profoundly influenced by the continued evolution of media technologies and the strategic engagement of athletes and organizers.

3.2 Youth Culture

Youth culture is deeply intertwined with extreme sports; a symbiotic relationship where each fuels the other's evolution. The rebellious spirit, the desire for self-expression beyond societal norms, and the inherent risk-taking that defines many extreme sports perfectly mirrors the often-chaotic energy of youth. This isn't simply about mimicking; it's a shared language, a visual and visceral representation of a generation's aspirations and anxieties. Think of skateboarding in the 1970s and 80s – a movement born on the streets, a rejection of mainstream culture expressed through creative, gravity-defying maneuvers. This rebellious energy resonated deeply with a youth generation seeking an alternative voice.

The influence extends beyond simple participation. Extreme sports heavily influence fashion trends; think of the baggy pants and loose-fitting clothing initially adopted for skateboarding and snowboarding, then adopted into wider youth subcultures. Music too, reflects this connection. The aggressive rhythms and energy of punk rock found a natural home in the soundtrack of extreme sports, further cementing the rebellious ethos. Even film and television have mirrored this connection, often portraying extreme sports athletes as individualistic figures challenging authority, and defying expectations, becoming cultural icons in their own right. Their narratives resonate powerfully with young people seeking their own path.

However, the relationship isn't always harmonious. Commercialization and the mainstreaming of extreme sports have, in some ways, diluted the original rebellious spirit. The sponsorship deals, the carefully crafted image, sometimes clash with the authentic, raw energy that initially captivated youth culture. Yet, this commercialization also provides the financial fuel for new innovations, and further exposure for the sports themselves, potentially attracting a new generation of participants and continuing the cyclical relationship. The tension between authenticity and commercialism is a constant negotiation within extreme sports and youth culture, a constantly evolving dynamic.

The media's role is crucial in shaping this connection. Television coverage, social media, and online platforms have amplified the visibility and influence of extreme sports. Viral videos showcasing breathtaking feats and near-misses create a spectacle that captivates a global youth audience. The accessibility of this media further normalizes and celebrates risk-taking behaviors, and alternative lifestyles, allowing youths to connect with like-minded individuals across geographical boundaries. This global community fosters a sense of belonging and shared identity, a crucial element in youth development and self-discovery.

Furthermore, technological advancements are integral to this evolution. High-definition cameras, drones, and sophisticated editing techniques create visually stunning content, further enhancing the appeal of extreme sports. This technology not only facilitates the sharing of these experiences, but it also directly contributes to the progression of the sports themselves. Improved equipment, more precise measurements, better safety gear, all result from technological innovation fueled by the ever-growing demand for both pushing limits and increased safety. This symbiotic relationship ensures the continued growth and evolution of extreme sports and their inherent link to youth culture. The future of this relationship will no doubt continue to be shaped by technological innovations, continuing this constant feedback loop.

The connection between youth culture and extreme sports extends far beyond mere trends. It speaks to a deeper desire for self-expression, a rejection of conformity, and an embrace of challenge. The inherent risks, the constant push for personal boundaries, reflects the uncertainties and explorations that define adolescence and the search for one's identity. Extreme sports offer a tangible expression of these anxieties and ambitions, a canvas for youth to paint their own narrative. Understanding this dynamic helps us appreciate the complex interplay between these seemingly disparate forces, and helps to understand how extreme sports have fundamentally shaped and continue to shape modern youth culture. This is more than just a trend; it's a powerful reflection of our times and a window into the evolving landscape of human experience and the youth's role within it. The enduring link will likely continue to define generations to come.

3.3 Impact on Society

Extreme sports' societal impact extends far beyond the athletic arena. The sheer spectacle of daring feats, broadcast globally through sophisticated media platforms, has profoundly shaped popular culture. From fashion trends mimicking the edgy aesthetics of skateboarding and snowboarding to the influence on music and film, extreme sports have become a potent cultural force, reflecting and shaping societal values around risk-taking, individuality, and pushing boundaries.
This influence on youth culture is particularly significant. Young people, often drawn to the rebellious spirit and self-expression inherent in extreme sports, find role models and inspiration in athletes who defy convention and embrace challenges. The visual storytelling of these athletes, often amplified through social media, fosters a sense of community and shared identity among a global youth population. This shared identity, fueled by online interaction and shared experiences,

transcends geographical boundaries, fostering a global dialogue around adventure, risk, and personal achievement.

Furthermore, the economic impact is undeniable. The global extreme sports industry generates significant revenue through sponsorships, merchandise sales, media rights, and tourism. The creation of specialized equipment, from high-performance apparel to innovative safety gear, fuels technological advancements in materials science and engineering, fostering innovation in fields beyond the immediate scope of sports itself. These industries, in turn, create jobs and stimulate economic growth in local communities that host major events or attract extreme sports enthusiasts.

However, the societal impact is not without its complexities. The inherent risk in these sports raises concerns about safety and injury prevention. The pressure on athletes to achieve increasingly audacious feats can lead to burnout, injury, and even fatal accidents. These risks necessitate a constant reassessment of safety standards and regulations, a process that involves collaboration between athletes, governing bodies, and equipment manufacturers. A balanced approach is crucial; one that celebrates the exhilarating aspects of extreme sports while mitigating the dangers and promoting responsible participation.

Moreover, the media's portrayal of extreme sports can sometimes be overly sensationalized, potentially glamorizing risky behavior and neglecting the rigorous training and discipline required. This necessitates a critical analysis of media representations, promoting a more nuanced and responsible portrayal of these athletes' dedication, perseverance, and the ethical considerations inherent in their pursuits. Responsible reporting should highlight the mental fortitude, dedication to training, and risk management strategies employed by athletes, offering viewers a balanced perspective and avoiding the glorification of reckless actions.

Finally, extreme sports present both opportunities and challenges concerning environmental sustainability. While the pursuit of these activities frequently takes place in pristine natural environments, there are inherent environmental concerns, from habitat disruption to waste management. The industry must proactively address these issues, adopting sustainable practices in equipment manufacturing, event organization, and overall environmental stewardship. The emphasis on responsible environmental practices is not merely a matter of compliance but a fundamental aspect of preserving the very environments that nurture and inspire these thrilling pursuits. The long-term sustainability of extreme sports is directly linked to the responsible management and preservation of the natural world.

3.4 Technological Advancements

Technological advancements have fundamentally reshaped the landscape of extreme sports, propelling athletes to previously unimaginable heights and pushing the boundaries of human potential. From the materials used in equipment to the methods of capturing and broadcasting events, innovation has been the driving force behind the explosive growth and evolution of these disciplines. Consider the radical changes in snowboarding, for instance. Early boards were rudimentary, often homemade, and lacked the sophisticated design and materials of today's high-performance models. Now, advanced composites and innovative shapes allow for unparalleled control, speed, and maneuverability, leading to more daring tricks and breathtaking feats. This evolution mirrors advancements across other extreme sports, demonstrating a consistent pattern of technological influence.

The integration of data analysis and wearable technology represents another significant leap forward. Sophisticated sensors embedded in equipment, and even directly on athletes' bodies, collect real-time performance data, offering unprecedented insights into biomechanics, speed, and other key metrics. This data empowers athletes to optimize their training, refine their techniques, and ultimately achieve new levels of proficiency. Coaches can utilize this information to create personalized training plans that push individual limits safely, minimizing the risk of injury while maximizing athletic growth. Furthermore, this technology contributes to a deeper understanding of the physiological demands of these high-intensity activities, leading to breakthroughs in injury prevention and rehabilitation.

Live streaming and virtual reality (VR) have revolutionized the way audiences engage with extreme sports. High-definition cameras capture breathtaking footage from every angle, offering viewers immersive experiences that were previously unthinkable. VR technology allows spectators to feel as though they're right there on the mountain, in the air, or on the wave, significantly enhancing the engagement and excitement surrounding the events. This level of access has broadened the appeal of extreme sports to a much wider audience, fostering a global community of enthusiasts and fueling the growth of the industry. The immediacy and accessibility of these technologies have also made it easier for aspiring athletes to connect with mentors and role models, accelerating the learning curve for those seeking to participate in extreme sports.

However, the rapid pace of technological advancement in extreme sports raises important ethical considerations. The use of performance-enhancing technologies, while sometimes debated, requires careful monitoring and regulation to ensure fairness and safety. The potential for misuse or the development of unequal access to advanced technologies could create an uneven playing field, undermining the principles of competition and sportsmanship. Striking a balance between embracing innovation and maintaining the integrity of the sport remains a critical challenge for governing bodies and the athletes themselves. This balancing act demands ongoing dialogue and collaborative efforts to establish fair guidelines and ensure responsible use of emerging technologies.

Finally, considering the environmental impact of extreme sports is vital as technology advances. Sustainable practices in the manufacturing of equipment and the organization of events are increasingly important, as the industry's footprint grows. The use of recycled materials, energy-efficient transport, and waste-reduction initiatives all contribute to mitigating the environmental effects of increased participation and media coverage. Furthermore, technological solutions can assist in monitoring and managing the environmental impacts of these sports, ensuring that their growth doesn't come at the expense of the planet's resources. The future of extreme sports hinges on a commitment to sustainability and responsible innovation, ensuring that the thrill of the experience doesn't overshadow the importance of environmental stewardship.

3.5 Future Trends

Virtual reality (VR) and augmented reality (AR) technologies are poised to revolutionize the experience and accessibility of extreme sports. Imagine training for a snowboarding competition using a VR simulator that perfectly replicates the feel of the slopes, or learning to surf using AR overlays that guide your movements and provide real-time feedback. This immersive technology will not only enhance training but also allow individuals with physical limitations to participate in extreme sports in ways previously unimaginable. The precision and realism of these simulations offer a safe and controlled environment to hone skills, reducing the risk of injury and accelerating the learning process. Furthermore, the integration of biometric sensors within VR and AR systems provides athletes with valuable data on their performance, enabling personalized training programs and further pushing the boundaries of human potential. We will likely see a significant increase in both participation and spectator engagement as this technology matures.

The rise of data analytics and wearable technology will redefine performance optimization in extreme sports. Sophisticated sensors embedded in equipment and wearable devices gather immense amounts of data during performance. This detailed information—everything from heart rate and muscle exertion to speed and altitude—is analyzed using advanced algorithms to identify areas for improvement, optimize training strategies, and prevent injuries. This personalized approach to training fosters a more scientific understanding of athletic performance, moving away from intuition-based methodologies. Elite athletes will utilize this technology to refine their techniques and push the limits of what is physically possible, while amateur athletes can use it to improve their safety and enjoyment of the sport. Expect to see a much greater focus on personalized training plans that are tailored to the unique needs and goals of individual athletes.

Sustainability concerns will become increasingly central to the future of extreme sports. The environmental impact of travel, equipment production, and waste generation is undeniable. We can anticipate a growing emphasis on eco-friendly practices, from the use of sustainable materials in equipment manufacturing to the adoption of carbon-neutral travel solutions for competitions. Expect the emergence of sustainable tourism initiatives that minimize the environmental footprint of extreme sports events and promote responsible interaction with natural environments. This shift towards sustainability isn't just a matter of environmental responsibility; it's also about ensuring the long-term viability of extreme sports in the face of climate change and resource depletion. Organizations and athletes alike will need to adapt and prioritize sustainable practices if the sports are to flourish.

The lines between extreme sports and other forms of entertainment will continue to blur. We're already seeing the integration of extreme sports elements into video games, movies, and other media forms, blurring the boundaries between virtual and real experiences. This convergence will foster further creativity and innovation, creating new opportunities for engagement and participation. Expect to see increased investment in esports-style competitions based on extreme sports simulations, alongside the development of interactive entertainment experiences that let participants feel the adrenaline rush virtually. This intersection can broaden the appeal of extreme sports to a much wider audience, making them more accessible and engaging for individuals who might not otherwise participate. This crossover also opens up new avenues for sponsorship and revenue generation, supporting the growth and development of the industry.

Finally, the globalization of extreme sports will continue, fostering greater cultural exchange and competition. We can expect to see an increase in international competitions, with athletes from diverse backgrounds competing on a global stage. This increased interaction will lead to a richer, more varied landscape of sports styles and techniques, with athletes pushing boundaries not just physically but also culturally. The sharing of skills and techniques across cultures will likely result in innovative approaches to training, equipment design and the overall experience of participation. The growth in global participation also presents an opportunity for fostering understanding and cooperation between different cultures, promoting social cohesion and diversity within the world of extreme sports. The future of extreme sports is dynamic and exciting, driven by innovation and a desire to push the boundaries of human potential.

4 Chapter 4: Beyond the Physical

Success in extreme sports isn't solely about physical prowess; it's deeply rooted in the athlete's mental landscape. The ability to conquer fear, to push past perceived limitations, and to maintain unwavering focus under immense pressure—these are the pillars of mental fortitude, arguably more critical than raw strength or technical skill. Imagine the sheer terror of free-soloing a cliff face, the unwavering concentration required for a backflip on a mountain bike, or the split-second decisions made during a high-speed snowboarding run. These moments are not just tests of physical ability, but profound explorations of the human psyche.

Overcoming fear is a continuous process, not a single event. It's about understanding your fear, acknowledging its presence without letting it paralyze you. Elite athletes often employ visualization techniques, mentally rehearsing their routines repeatedly, building confidence and reducing anxiety. This mental preparation allows them to approach challenging situations with a sense of control and preparedness. They transform fear from a debilitating force into a source of motivation, transforming apprehension into anticipation. This mental mastery isn't innate; it's meticulously cultivated through dedicated training and self-awareness.

The mental game extends beyond simply conquering fear; it encompasses unwavering discipline and relentless dedication. This is where mental resilience plays a vital role, shaping the athlete's ability to bounce back from setbacks, to learn from failures, and to maintain a positive mindset even in the face of adversity. Extreme sports are inherently risky; falls, injuries, and failures are inevitable. Athletes who can analyze their mistakes, adapt their strategies, and maintain a growth mindset are far more likely to achieve long-term success. Their ability to learn from each experience, to constantly refine their techniques, and to remain focused on their goals distinguishes them from those who succumb to self-doubt or despair. This is where the true grit and mental fortitude shine, shaping not only their sporting achievements, but their entire lives.

Discipline manifests in countless ways, from meticulous training schedules and nutritional plans to the consistent pursuit of self-improvement. It's about making sacrifices, prioritizing training over distractions, and consistently pushing the boundaries of comfort. It's about the countless hours spent honing skills, the early morning workouts, the unwavering dedication to physical and mental conditioning. This unwavering commitment reflects not only their passion for their sport but also a broader dedication to self-discipline and achieving their full potential. A champion's attitude demands consistent work ethic and relentless commitment to self-improvement. The mental aspect is paramount, allowing the athlete to endure relentless physical demands, push through pain, and maintain focus even when exhaustion sets in.

Beyond individual psychological strategies, the role of coaching and support networks is crucial in fostering mental resilience. A skilled coach acts as a mentor, providing guidance, feedback, and support throughout the athlete's journey. They help identify areas for improvement, offer strategies for overcoming mental barriers, and provide a stable source of encouragement during challenging moments. Moreover, the close-knit communities found in many extreme sports provide a network of peers, fostering mutual support and shared experiences. This collective support system creates an environment where athletes can share their triumphs and struggles, offering each other encouragement and understanding, reinforcing the importance of a robust psychological strategy.

The mental preparation extends beyond individual training; it also impacts the athlete's overall well-being. Learning to manage stress, maintain emotional balance, and cultivate self-awareness are vital components of success. The pressures of competition, the risks inherent in the sport, and the intense focus required can take a toll on mental health. Athletes who prioritize self-care, engage in mindfulness practices, or seek professional support demonstrate a profound understanding of the mental game. This balance between physical and mental well-being is essential for sustaining a successful career in extreme sports, demonstrating the critical role of mental fortitude in achieving peak performance. The pursuit of excellence in extreme sports is not merely a physical endeavor but a testament to the enduring power of the human mind.

4.1 Mental Fortitude

Mental fortitude isn't some mystical power; it's a skill honed through deliberate practice and self-awareness. Think of it like training for a marathon – you wouldn't expect to run 26 miles without consistent preparation. Similarly, facing the immense physical and mental challenges of extreme sports requires cultivating an unwavering inner strength. This involves understanding your personal strengths and weaknesses, acknowledging your fears, and developing coping mechanisms to manage pressure and setbacks. It's about building a resilient mindset that can bounce back from failure, learn from mistakes, and persistently pursue your goals, even when the odds seem insurmountable.

Building this resilience starts with self-reflection. What are your triggers? What situations typically lead to doubt or anxiety? Identifying these patterns is the first step towards managing them. Extreme athletes often use journaling or mindfulness techniques to process their emotions and develop self-awareness. This isn't about suppressing negative feelings; it's about understanding their root cause and developing healthy strategies for dealing with them. For example, recognizing that a fear of failure stems from past experiences allows you to reframe your perspective and approach challenges with a newfound confidence.

Visualization is another powerful tool. Elite athletes frequently use mental imagery to prepare for competition. They mentally rehearse their performance, visualizing every movement, anticipating potential problems, and mentally reinforcing their success. This process trains the mind to react effectively under pressure and builds confidence by familiarizing itself with the demands of the sport. Imagine yourself flawlessly executing a difficult maneuver; feel the thrill of success. This mental rehearsal strengthens neural pathways and prepares your mind for the physical execution.

Developing mental toughness is not a passive process; it's an active and continuous commitment to self-improvement. It means embracing discomfort, pushing past your perceived limits, and celebrating small victories along the way. Think of each training session, each competition, as an opportunity to grow mentally and physically. Every time you confront your fears and overcome a challenge, you build your mental fortitude, reinforcing your self-belief and resilience. This process is cumulative; each success strengthens your ability to face future challenges with greater confidence and determination.

Overcoming setbacks is an inevitable part of the journey. In extreme sports, falls, injuries, and failures are commonplace.

However, it's how you respond to these setbacks that truly defines your mental fortitude. Instead of viewing them as defeats, reframe them as learning opportunities. Analyze what went wrong, identify areas for improvement, and adjust your strategy accordingly. This mindset fosters resilience and prevents setbacks from derailing your progress. Remember the lessons learned, and use them to fuel your future efforts. Your response to failure is a key indicator of your mental strength.

Beyond individual strategies, support systems play a critical role in cultivating mental fortitude. Surrounding yourself with positive influences – coaches, teammates, family, and friends – can significantly impact your mental well-being. A strong support network provides encouragement, guidance, and a safe space to process challenges and celebrate achievements. Open communication about your struggles and successes is crucial; sharing your experiences can reduce feelings of isolation and promote a sense of community. Lean on your support network for guidance and encouragement when you face adversity.

Ultimately, mental fortitude in extreme sports, and indeed in life, is about embracing the journey, celebrating progress, and learning from setbacks. It's a continuous process of self-discovery, growth, and resilience. By actively practicing self-reflection, visualization, and developing a strong support system, you can cultivate an unwavering inner strength that empowers you to conquer challenges and achieve your full potential. The path is not always easy, but the rewards of mental fortitude are immeasurable, both within the world of extreme sports and beyond. Remember, your mental strength is just as crucial as your physical ability; nurture it, and it will serve you well throughout your life.

4.2 Overcoming Fear

Fear. It's a primal instinct, a hardwired response designed to protect us. In extreme sports, however, fear becomes a formidable opponent, a wall between the athlete and the exhilarating precipice of achievement. Conquering this fear isn't about its eradication; it's about understanding, managing, and ultimately, using it to your advantage.

Many athletes describe fear as a physical sensation, a tightening in the chest, a tremor in the hands. It's a visceral experience that can paralyze, leading to hesitation and ultimately, failure. But this same fear, when properly channeled, can transform into a potent source of focus. The adrenaline it produces sharpens reflexes, enhances concentration, and fuels the drive to succeed. The key lies not in ignoring the fear, but in learning to control the physical manifestations and harness its energy.

Think of it like this: your body is signaling danger, alerting you to potential risk. Instead of fighting this signal, acknowledge it. Breathe deeply, feel the tension, and consciously relax your muscles. Visualize your routine, replaying successful runs in your mind's eye. This mental rehearsal calms your nervous system, reducing the physical impact of fear. It creates a mental anchor, a place of calm amidst the storm of adrenaline.

Progressive exposure is another powerful tool. Start small. Don't jump straight into the most challenging aspect of your sport. Gradually introduce yourself to fear-inducing elements, one step at a time. This incremental approach allows you to build confidence and resilience. Each successful step forwards reinforces your belief in your ability to manage and overcome fear. This methodical progression builds self-efficacy, a crucial component of success in high-stakes activities.

Beyond physical and mental techniques, the role of a strong support system cannot be overstated. Surround yourself with people who understand your ambitions, who provide encouragement and reassurance. A coach, a teammate, a family member—these individuals can act as anchors, bolstering your confidence and reminding you of your capabilities. Sharing your fears can also help to demystify them, reducing their overwhelming power. Open communication fosters trust and reduces feelings of isolation, a common experience for athletes facing challenging feats.

Preparation plays a crucial part in minimizing fear. Thorough knowledge of the terrain, the equipment, and the potential risks significantly reduces the element of surprise. Rigorous training reduces the margin for error and builds a foundation of confidence that transcends individual attempts. Mastering fundamental skills reduces anxiety linked to uncertainty, allowing the athlete to focus on the task at hand rather than dwelling on the potential for failure. It is important to understand the inherent risks associated with extreme sports.

Ultimately, conquering fear in extreme sports is a continuous journey, a process of learning and growth. It's about developing a sophisticated relationship with fear, learning to recognize and understand its signals, and utilizing its energy to drive performance. It's about pushing your limits in a controlled and calculated manner, never underestimating the dangers, but always recognizing the strength, resilience, and potential that resides within you. This journey will demand self-awareness, constant learning and adaptation, as well as the wisdom to know when to push boundaries and when to retreat. It's a journey that will constantly challenge you. Embrace the challenge; for within the realm of controlled risk lies the path to achieving extraordinary accomplishments and the potential for unprecedented personal growth. Remember, the most exhilarating experiences often involve a certain degree of controlled fear. It's in confronting these fears, and mastering the techniques to manage them that you unlock your true potential as an athlete, and as an individual. The line between fear and excitement can be blurred, the art is in learning to leverage the power of fear for the greater cause of success. This constant negotiation with your own fear is a core element in the extreme athlete's psyche and ultimately what defines the character of those pushing the boundaries of human potential.

5 Chapter 5: Innovation in Design

The relentless pursuit of bigger, faster, higher in extreme sports has fueled an equally relentless drive for innovative design. This isn't just about making things look cool; it's about survival. Consider the evolution of the snowboard. Early boards were rudimentary, essentially planks of wood. Today, we have high-tech composites, carefully engineered shapes, and bindings offering unparalleled control and safety. This evolution reflects not only advancements in materials science but also a deep understanding of biomechanics and the forces involved in high-speed maneuvers. The design process itself has become a crucial component, with extensive testing and feedback loops shaping the final product.

Improved safety measures are paramount. The development of helmets, protective pads, and specialized apparel has significantly reduced injury rates. These aren't just simple additions; they're sophisticated pieces of engineering designed to absorb and disperse impact forces, protecting vulnerable areas of the body. The incorporation of advanced materials like Kevlar and carbon fiber has allowed for lighter, stronger, and more flexible protective gear, enhancing both safety and performance. The ongoing research into impact dynamics and injury prevention continues to refine these designs, making extreme sports safer for participants.

Sustainability is increasingly integral to design thinking within extreme sports. The industry is facing growing pressure to minimize its environmental footprint. Manufacturers are actively seeking sustainable materials, reducing waste in production processes, and extending the lifespan of equipment through durability and repairability. This shift toward eco-conscious practices includes exploring bio-based materials, recycling programs, and the use of more efficient manufacturing techniques. This isn't just a trend; it's a necessary adaptation for an industry that often takes place in fragile natural environments. The shift toward sustainability not only reduces the negative impacts of extreme sports but also creates opportunities for innovation and economic diversification within the industry.

Beyond the physical equipment, the development of innovative tracking and monitoring systems is revolutionizing the athlete experience. GPS tracking, wearable sensors, and data analytics provide athletes with real-time feedback on their performance, enabling precise adjustments to technique and training regimens. This technology also facilitates safety protocols, allowing for rapid response in emergencies. Moreover, data analysis helps designers understand how equipment performs under stress, leading to further refinements and innovations. The interconnectedness of these data systems with other areas of training, coaching, and safety protocols has led to significant leaps forward.

The design of extreme sport equipment isn't a static process; it's a continuous evolution driven by athletes' demands, technological advancements, and a growing awareness of environmental responsibilities. Each new innovation, from the most minor adjustment to a major technological leap, represents a step toward pushing the boundaries of human potential while minimizing risk and environmental impact. The integration of safety, performance enhancement, and ecological consciousness marks a significant transition within the industry, showcasing how extreme sports can inspire both exceptional feats of athleticism and environmentally responsible design practices. This innovative spirit isn't just about the thrill of competition; it's a testament to human ingenuity and the relentless pursuit of progress. The future of extreme sports rests not just on the athleticism of its participants, but on the creativity and ingenuity of its designers. The constant striving for improvement in equipment design will undoubtedly shape the future of the sport for generations to come, ensuring its continued evolution. The iterative process of testing, refining, and innovating is at the very heart of progress, constantly pushing boundaries and pushing the evolution of safety and sustainability further into the future. The very essence of extreme sports – the constant pursuit of pushing limits – manifests itself in this unrelenting drive towards better, safer, and more sustainable equipment. Each piece of equipment, each innovation, reflects a journey – a journey of technological advancement, a journey towards better safety, and a journey of environmental responsibility.

5.1 Equipment Evolution

The evolution of equipment in extreme sports isn't just about incremental improvements; it's a continuous revolution driven by the relentless pursuit of performance and safety. Consider snowboarding, for example. Early boards were rudimentary, essentially planks of wood. Today, we have high-tech composites, sophisticated bindings offering precise control and responsiveness, and boots designed for optimal power transfer and comfort, all meticulously engineered to minimize injury risk while maximizing performance potential. This isn't merely about faster speeds; it's about enabling athletes to execute increasingly complex and daring maneuvers with a greater margin of safety.

This evolution isn't confined to snowboarding. Similarly, in skateboarding, the transition from simple wooden boards to high-performance decks crafted from multiple layers of wood and advanced materials represents a significant leap forward. The development of specialized trucks and wheels, designed for different terrains and riding styles, has dramatically expanded the possibilities and technical skills in the sport, creating avenues for innovation in trick design and execution previously unimaginable. The improved grip and durability of these modern components have also substantially reduced injuries.

Furthermore, advancements in protective gear have played a crucial role in the growth of these sports. Helmets, once basic and clunky, are now lightweight, aerodynamic, and engineered to absorb impacts with unprecedented effectiveness. Similarly, knee pads, elbow pads, and other protective gear have progressed from bulky, uncomfortable garments to sophisticated, form-fitting designs that minimize restriction while maximizing protection. This enhanced safety has empowered athletes to push their limits further, leading to more ambitious and innovative performances.

The interplay between material science and extreme sports design is particularly fascinating. The introduction of carbon fiber, for instance, has revolutionized the design of many extreme sports products. Its lightweight yet incredibly strong nature allows for the creation of lighter, more responsive equipment, leading to improved performance and agility. This is evident in mountain biking, where carbon fiber frames have become increasingly prevalent, contributing to bikes that are both lighter and more durable, capable of withstanding the rigors of extreme terrain.

Beyond materials, the integration of advanced technologies continues to refine equipment design. For instance, the use of sensors and data analysis in equipment development is revolutionizing the way athletes train and perform. By collecting real-time data on forces, speeds, and other performance metrics, engineers can refine designs to optimize performance and minimize stress on the athlete's body. In climbing, for instance, the use of smart climbing shoes, incorporating sensors to monitor pressure distribution and strain, is improving not only the performance but also the safety of the sport.

Sustainability is also becoming a key factor in equipment evolution. There's a growing demand for eco-friendly materials and manufacturing processes, leading to innovative designs that prioritize both performance and environmental responsibility. Companies are increasingly using recycled materials and exploring biodegradable alternatives. This shift reflects a broader societal concern for environmental impact, and the extreme sports industry is actively embracing more sustainable practices, recognizing their role in protecting the environments in which they operate.

The evolution of equipment in extreme sports is an ongoing process, driven by a symbiotic relationship between athletes, engineers, and materials scientists. As athletes push the boundaries of what's physically possible, engineers and scientists respond by creating equipment that enables them to achieve their goals while minimizing risks. This continuous cycle of innovation has not only transformed the landscape of extreme sports, but also led to significant advancements in materials science, engineering, and design principles, pushing the limits of what we know to be possible in engineering materials and design. This constant push for improved performance and increased safety will continue to shape the future of extreme sports for generations to come, setting new standards for human potential and technological advancements.

5.2 Safety and Technology

The relentless pursuit of pushing limits in extreme sports necessitates a parallel advancement in safety technology. This isn't just about padding and helmets; it's about a holistic approach integrating cutting-edge materials science, biomechanics, and data analysis to minimize risk while maximizing the thrill. Think lightweight, high-strength composites replacing traditional materials, offering superior protection with reduced weight, thus enhancing athletic performance.

Consider, for example, the evolution of snowboarding equipment. Early boards were rudimentary, leading to countless injuries. Today, advanced materials and designs—from rockered profiles to specialized bindings—dramatically reduce the risk of leg and ankle injuries. This is a testament to the iterative process of design refinement driven by the need for both performance and safety.

Moreover, technological innovations extend beyond the equipment itself. Sophisticated monitoring systems, like wearable sensors embedded in protective gear, provide real-time data on an athlete's vital signs, impact forces, and biomechanics. This information isn't just for post-event analysis; it allows for immediate adjustments in training and technique, minimizing the chance of injury before it happens. Think of it as a proactive safety net, continuously adapting to the athlete's performance.

Furthermore, the use of sophisticated simulations and virtual reality training is gaining traction. These technologies allow athletes to practice complex maneuvers and high-risk situations in a controlled environment, gradually building their skills and confidence without the immediate physical danger. This is analogous to flight simulators for pilots, a crucial element in aviation safety training. The potential for injury reduction is enormous, particularly in extreme sports with inherently high risk profiles, such as BASE jumping or big-wave surfing.

Beyond individual athlete safety, technological advancements are enhancing overall safety within the extreme sports environment. For instance, improved weather forecasting models and predictive analytics can help event organizers make more informed decisions, reducing risks associated with unpredictable conditions. Similarly, drone technology offers real-time monitoring of remote locations, facilitating swift emergency responses and enhancing overall safety protocols. This is particularly relevant to mountainous or remote locations frequently used in certain extreme sports.

It's crucial to recognize that the pursuit of technological innovation should never come at the expense of the inherent risks of extreme sports. The inherent thrill and challenge remain central to their appeal. However, by implementing these technological advancements, we can create a safer environment that allows athletes to push their boundaries while minimizing the potential for catastrophic injury. The aim is not to eliminate risk entirely, which would negate the very essence of these sports, but to mitigate it through a responsible and balanced approach. The challenge lies in creating a harmonious balance between pushing limits and ensuring the well-being of participants. The integration of technology should enhance, not replace, the intrinsic skills and judgment athletes cultivate. The future of extreme sports hinges not only on the continuous evolution of athletic skill and creative expression but also on the concurrent development of innovative safety technologies. By viewing safety not as a constraint but as a catalyst for innovation, we unlock new possibilities while ensuring that the exhilaration of these sports remains accessible to dedicated athletes for generations to come. This collaborative effort, marrying human skill with technological advancement, is not merely about injury prevention; it's about enabling the continued evolution and expansion of extreme sports—a vital and dynamic sector of human exploration.

5.3 Sustainable Practices

The environmental footprint of extreme sports, often overlooked amidst the thrill and spectacle, demands careful consideration. From the manufacturing of specialized equipment to the travel involved in competitions, the industry generates a significant impact. However, a growing awareness is driving a shift towards more sustainable practices, recognizing that responsible environmental stewardship is crucial for the long-term health of the sport. This isn't merely about reducing carbon emissions; it's about safeguarding the very landscapes that provide the backdrop for these exhilarating activities.

Sustainable manufacturing is paramount. The creation of high-performance equipment, often involving resource-intensive materials like carbon fiber and various plastics, needs to evolve. Innovations in materials science are offering exciting possibilities, with bio-based composites and recycled materials entering the picture. Companies are increasingly focusing on reducing waste throughout their production processes, implementing circular economy models that prioritize reuse and recycling. This is not just a trend; it's a necessity if extreme sports are to remain viable in a world increasingly concerned about resource depletion.

Beyond the equipment itself, travel associated with competitions and training presents a substantial challenge. Athletes and their support teams frequently travel long distances, contributing significantly to carbon emissions. While completely eliminating this travel may be impractical, strategies exist to minimize its impact. Choosing more fuel-efficient modes of transport, optimizing travel schedules to reduce journeys, and prioritizing local or regional events whenever possible are all practical steps that can make a tangible difference.

Beyond minimizing the environmental impact, supporting local communities is essential for long-term sustainability. Extreme sports often take place in remote areas, which may lack infrastructure and economic opportunities. By ensuring that local communities benefit from the economic activity generated by these events, we build a mutual relationship that fosters both economic and environmental sustainability. This could involve employing local guides, sourcing supplies locally, and investing in community development projects. The goal is not just to enjoy the thrills of the sport, but also to contribute positively to the environment and people who make it possible.

The choice of locations for competitions and training needs careful assessment. Protecting delicate ecosystems is crucial. Organizing events in areas already equipped to handle the influx of people, using existing infrastructure to reduce the environmental footprint, and limiting the overall impact on the natural environment are vital. It's a balance between showcasing the beauty of these landscapes and protecting them from degradation. Sustainable practices must be a fundamental consideration in the planning and execution of any event.

Furthermore, athletes themselves can play a significant role in promoting sustainable practices. By adopting eco-conscious lifestyles, supporting sustainable businesses, and advocating for environmentally responsible policies, they can influence their fan base and inspire positive change. This could encompass choices like using reusable water bottles, reducing waste, opting for sustainable travel, and championing sustainable brands. Such actions send a powerful message and encourage others to adopt similar practices.

Education and awareness play a crucial role. Increased awareness about the environmental impact of extreme sports can motivate both individuals and organizations to adopt more sustainable practices. Educational programs and initiatives can highlight the importance of environmental stewardship, offering concrete examples of sustainable alternatives. Raising awareness about the choices available and their potential impact can make a significant difference in fostering widespread adoption of eco-friendly practices.

Ultimately, the sustainability of extreme sports depends on the collective effort of athletes, manufacturers, organizers, and fans. By embracing sustainable practices, we ensure that future generations will continue to enjoy the thrill and excitement of extreme sports without compromising the health of our planet. This is not simply a matter of compliance, it's a commitment to ensuring the enduring future of the sport and the preservation of the environment we so passionately utilize. The continued vibrancy of extreme sports necessitates a transition towards sustainability; it's a responsibility we must collectively embrace.

6 Chapter 6: Global Impact

Extreme sports have transcended national borders, becoming a truly global phenomenon. International competitions, like the X Games and the Olympics' inclusion of snowboarding and skateboarding, serve as powerful platforms for showcasing athletic prowess and fostering cultural exchange. Athletes from diverse backgrounds compete, sharing their skills and inspiring each other, creating a melting pot of talent and camaraderie rarely seen in other sports. This globalized competition isn't just about medals; it's about building bridges between cultures and promoting understanding on a worldwide scale. The shared adrenaline and risk-taking create a unique bond between athletes, regardless of their origin.

The economic impact of extreme sports is substantial and multifaceted. Major events generate significant revenue through tourism, sponsorship, media rights, and merchandise sales. Local communities hosting competitions often experience economic boosts, with increased hotel bookings, restaurant patronage, and employment opportunities. Moreover, the industry supports a large network of equipment manufacturers, designers, and retailers, creating jobs and stimulating economic growth on a global scale. The rise of extreme sports tourism has also had a transformative effect on previously remote locations, attracting investment and infrastructure development.

However, the rapid growth of extreme sports isn't without its environmental concerns. Many extreme sports, by their very nature, take place in pristine natural environments. The increasing number of participants and events can lead to habitat degradation, pollution, and the disruption of fragile ecosystems. For instance, snowboarding and skiing can cause damage to alpine meadows, while mountain biking can erode trails and disturb wildlife. Sustainable practices, such as responsible trail building, waste management, and limiting access to sensitive areas, are crucial for mitigating these negative impacts and ensuring the long-term viability of the sport.

Beyond the economic aspects and environmental challenges, extreme sports play a significant role in community building. Local extreme sports clubs and organizations often foster a strong sense of belonging and shared identity among participants. These communities provide support networks, training opportunities, and a platform for sharing knowledge and experience. This collaborative spirit extends beyond individual sports, often uniting people from diverse backgrounds through a common passion for pushing boundaries. Such communities can also facilitate positive social change, inspiring environmental awareness, promoting inclusivity, and supporting local initiatives. Many extreme sports organizations actively engage in local community outreach projects.

The global impact of extreme sports is complex and multifaceted. While international competitions drive economic growth and cultural exchange, environmental sustainability and responsible community development remain crucial considerations for the future. It's a narrative of success, shared experience, and mutual respect – yet one that demands careful management to ensure its continued positive global contribution. The balance between promoting the thrill and adventure of extreme sports and safeguarding the environment and local communities will determine the long-term legacy of this powerful global phenomenon. Successfully navigating this challenge will require collaborative efforts from athletes, organizers, governing bodies, and environmental advocates, working together to ensure a sustainable and equitable future for extreme sports worldwide. Ultimately, the continued growth and success of extreme sports hinge upon a commitment to responsible practices that minimize negative environmental impacts while maximizing positive societal contributions. The global community must embrace a balanced approach, acknowledging the excitement and innovation of these sports while simultaneously protecting the planet and fostering vibrant, inclusive communities.

6.1 International Competitions

International competitions are the crucible where athletes forge their legacies, not just through individual triumphs but also through the powerful exchange of skills and cultures. These events aren't simply sporting contests; they're vibrant showcases of human potential, pushing the boundaries of what's physically and mentally possible. The energy is palpable, a shared adrenaline rush felt by competitors and spectators alike.

Consider the sheer logistical undertaking behind staging these events. From securing sponsorships and navigating international regulations to coordinating travel and accommodation for athletes from diverse backgrounds, the organizers face an incredible challenge. This complexity, in itself, highlights the global reach and impact of extreme sports. The meticulous planning, risk assessment, and meticulous attention to detail underscore the seriousness and importance of these competitive arenas. The success of these events hinges on effective collaboration, precise execution, and the seamless integration of various logistical elements.

The economic impact of these global competitions is significant. Tourism revenue surges as fans flock to witness the spectacle, boosting local economies and creating employment opportunities. Sponsors reap the benefits of global brand exposure, associating themselves with the thrill and excitement of the events. Local businesses, from hotels to restaurants, feel the positive ripple effects, highlighting the economic vitality of these international gatherings. This positive financial impact is only amplified by extensive media coverage, bringing the excitement to a global audience and further boosting the appeal of extreme sports on the world stage.

Beyond the economic aspects, there's an undeniable cultural exchange that occurs. Athletes, coaches, and fans from all corners of the globe come together, sharing traditions, perspectives, and experiences. These interactions foster mutual understanding, breaking down cultural barriers and promoting global camaraderie. This intercultural exchange enriches not only the participants but also enriches the global community that enjoys and participates in extreme sports. The sharing of perspectives and knowledge contributes to the universal appreciation of the skill, dedication and athleticism showcased in this increasingly popular sport.

However, the global nature of these competitions also presents challenges. Environmental concerns are paramount; the ecological impact of hosting large-scale events needs careful consideration and responsible mitigation strategies. Sustainable practices, from minimizing waste to promoting eco-friendly transportation, are increasingly important to reduce the environmental footprint of these internationally impactful events. Responsible environmental stewardship is critical for the long-term sustainability of the sport.

Furthermore, ensuring fair competition and equal opportunities for all participants requires diligent oversight. The rules and regulations must be transparent and consistently applied, regardless of nationality or background. This ensures a level playing field, promoting fair play and celebrating the accomplishments of athletes based on merit, skill, and dedication. This also contributes to building trust and integrity in a global competitive environment.

Ultimately, international competitions aren't just about winning or losing; they're about pushing limits, inspiring future generations, and forging connections across cultures. They showcase the power of human endeavor and the unifying spirit of competition. The legacy of these events transcends the immediate results, leaving a lasting impact on participants, communities, and the broader global landscape. The spirit of competition fosters unity, showcasing shared humanity in the pursuit of excellence. These events serve as powerful platforms for cultural exchange and inspire millions around the world. Their continued growth demonstrates the widespread appeal and global impact of extreme sports. They serve as powerful examples of global unity and the transformative potential of shared human experiences. The competitions demonstrate the enduring human passion for pushing limits and surpassing boundaries. The collaborative spirit fostered during these international events underscores the unifying power of extreme sports. The inspiring stories of resilience and triumph inspire generations, leaving a lasting legacy far beyond the competitive arena. They represent the culmination of years of dedication, training and the unwavering pursuit of excellence. The excitement, athleticism, and collaborative spirit at the heart of these events leave a lasting impression on all who participate and witness them, establishing these competitions as defining moments in the global extreme sports landscape.

6.2 Cultural Exchange

The global nature of extreme sports fosters a fascinating exchange of ideas and techniques. Athletes from diverse backgrounds bring unique perspectives, not only in their physical skills but also in their approaches to training, risk assessment, and even the very definition of what constitutes an "extreme" challenge. Consider the subtle differences in training philosophies between a Japanese snowboarder and a Chilean surfer—each shaped by their respective cultural contexts, their environments, and their ingrained approaches to risk-taking and community support. This cross-cultural pollination isn't limited to the athletes themselves. The design and manufacture of equipment, often driven by innovative solutions tailored to specific geographical and environmental factors, show a globalized exchange, with materials and technologies adopted and adapted across continents.

This isn't merely a sharing of practices; it's a dynamic process of adaptation and evolution. For instance, the techniques used by mountain climbers in the Himalayas have influenced climbing safety standards worldwide. Conversely, the development of certain types of climbing gear in Europe has spread rapidly to other regions, improving the safety and accessibility of climbing for individuals regardless of their geographical location or cultural background. The result is a constantly evolving set of best practices, pushing the boundaries of both skill and safety.

Moreover, the shared passion for these sports transcends language barriers and cultural differences. Competitions, both formal and informal, create a vibrant space for interaction and understanding. The shared language of athleticism, the common goal of pushing physical and mental limits, transcends cultural norms and fosters camaraderie amongst competitors from different backgrounds. This is not only witnessed in major international events but also in smaller, more localized gatherings where athletes connect and exchange knowledge.

The economic impact of this global exchange is substantial. Extreme sports generate revenue through tourism, media coverage, sponsorship, and the manufacture of equipment. This revenue isn't concentrated in a few dominant nations but rather distributed across the globe, supporting economies in diverse regions and providing opportunities in countries where the sports have taken root. This creates a powerful economic incentive for nations to support extreme sports' development within their borders.

However, the global reach of these sports also brings challenges. The environmental impact of extreme sports, particularly those involving significant travel or use of specialized equipment, cannot be ignored. The responsibility to minimize this environmental footprint falls on athletes, organizers, and the manufacturers of equipment. This includes embracing sustainable practices in manufacturing, minimizing travel, and promoting responsible environmental stewardship amongst participants. Balancing the growth and popularity of extreme sports with a commitment to environmental sustainability is a critical task.

Furthermore, it's crucial to address the potential for cultural appropriation and the need for respectful engagement. The adoption of techniques or styles from other cultures must be done with sensitivity and respect, avoiding the exploitation or misrepresentation of cultural heritage. It's important for athletes and organizations to engage in genuine dialogue and collaborations with local communities to ensure that the cultural exchange is mutually beneficial and respectful. Education and understanding are key to navigating this complex landscape and ensuring that the global growth of extreme sports doesn't come at the cost of cultural integrity. Ultimately, it is a balance between fostering global connections and maintaining the unique identities and heritages of each participating cul-

ture.
The future of extreme sports is inextricably linked to the continuation and enhancement of this cultural exchange. As more people from diverse backgrounds participate, the sports themselves will evolve, driven by new techniques, styles, and perspectives. This evolution will require not only athletic prowess but also a conscious effort to foster respectful collaboration and ensure the sustainability of the sports themselves and the environments in which they are practiced. The true success of extreme sports in the years to come will depend on its capacity to successfully navigate the complexities of cultural exchange, striking a balance between promoting globalization and preserving cultural uniqueness. This delicate balance is essential to ensure that the growth of the sports is sustainable, ethical, and enriches rather than diminishes the cultures from which it draws. The future success of extreme sports depends on the wisdom to manage this delicate balance, ensuring a future where the thrill of competition is matched by the respect and understanding that comes from cross-cultural collaboration. This understanding allows for the creation of a truly global community, sharing a common passion while respecting unique cultural identities.

6.3 Economic Influence

The economic impact of extreme sports extends far beyond the immediate revenue generated by ticket sales and sponsorships. It's a complex web of interconnected industries, creating a significant ripple effect across numerous sectors. Think of the manufacturing of specialized equipment—from high-tech skis and snowboards to advanced climbing gear and protective apparel—each piece demanding innovative design and materials, boosting related manufacturing jobs and technological advancements. This necessitates investment in research and development, pushing the boundaries of material science and engineering.

Beyond manufacturing, consider the tourism industry. Extreme sports events often draw large crowds, boosting local economies through increased hotel bookings, restaurant patronage, and demand for transportation services. This influx of tourism money revitalizes local communities, supporting businesses and creating employment opportunities. Many areas, once economically stagnant, have found renewed life by embracing extreme sports tourism, transforming themselves into attractive destinations. The economic boost is especially noticeable in smaller, more remote regions, often lacking other substantial economic drivers.

Media coverage further amplifies the economic influence. Television broadcasts, online streaming, and social media engagement all contribute to the immense popularity of extreme sports, attracting sponsors and creating lucrative advertising opportunities. This media attention, in turn, fuels the demand for extreme sports equipment and apparel, further stimulating manufacturing and retail industries. The professional athletes themselves, becoming global celebrities, generate substantial income through endorsements and sponsorships. These individuals, in essence, act as brand ambassadors, influencing consumer behavior and generating revenue streams for various companies.

Moreover, the economic influence extends to the development of infrastructure. Hosting major extreme sports competitions requires significant investment in venues and facilities, leading to long-term economic benefits for the hosting locations. The construction and maintenance of these facilities generate employment opportunities, and the ongoing operational expenses further contribute to local economies. Think of the specialized training facilities, the construction of dedicated parks and trails, or the improvement of existing infrastructure to accommodate the influx of participants and spectators.

These projects often stimulate further development in the surrounding areas.
The creation of specialized training centers is another crucial aspect of the economic landscape. These centers provide high-quality training for athletes of all skill levels, supporting the development of future champions while creating numerous jobs for coaches, trainers, and support staff. These centers also attract tourists seeking training or viewing opportunities, adding to the overall economic benefits. In essence, extreme sports generate a robust network of interconnected economic activities, contributing significantly to job creation, technological advancement, tourism revenue, and overall economic growth in various regions worldwide. Its sustained growth and evolving nature ensures that the economic influence will continue to expand in the coming years, presenting an engaging study in the dynamic relationship between sport, economics, and technological innovation.

6.4 Environmental Concerns

The environmental footprint of extreme sports is a complex issue, often overlooked amidst the thrill and spectacle. While the adrenaline-fueled pursuits inspire awe, they invariably leave a mark on the landscapes they utilize. From the pristine slopes consumed by skiers and snowboarders to the fragile coral reefs impacted by wave riders, the environmental cost demands our attention. This isn't about shaming athletes; instead, it's about acknowledging impacts and exploring solutions.
We must consider the transportation involved in reaching remote locations for competitions and training. Frequent flights, long drives using gas-guzzling vehicles, and the sheer logistics of moving equipment and personnel all contribute to significant carbon emissions. Furthermore, the construction of

infrastructure – ski resorts, skate parks, climbing walls – often necessitates deforestation and land alteration, disrupting delicate ecosystems.

The production of specialized equipment, while crucial for safety and performance, also generates considerable waste. Many materials used in the manufacture of boards, skis, helmets, and other gear are not easily recyclable, and the lifecycle of these products is often relatively short due to rapid technological advancements. This leads to significant amounts of discarded equipment ending up in landfills.

Mountain biking, for example, can cause trail erosion and habitat disturbance if not managed responsibly. The popularity of the sport, while beneficial for boosting economies of certain regions, needs careful regulation to prevent overuse and damage to trails. Similar concerns arise in other extreme sports like rock climbing and surfing. The constant need for "new frontiers" can lead to the unsustainable exploitation of natural environments.

However, the picture isn't entirely bleak. Increasingly, athletes, manufacturers, and governing bodies are recognizing their environmental responsibility. Sustainable practices are emerging. Manufacturers are experimenting with more eco-friendly materials like recycled plastics and sustainably sourced wood, reducing their carbon footprint. The use of renewable energy sources in the construction and maintenance of facilities is becoming increasingly common.

Initiatives promoting responsible tourism in extreme sports locations are also gaining traction. These programs focus on minimizing environmental impact through education, responsible waste management, and encouraging respect for the natural environment. Educating athletes about sustainable practices, encouraging them to leave no trace, and promoting responsible behaviour is crucial. They can become powerful advocates for environmental protection.

Furthermore, the inherent connection between extreme athletes and the natural world presents a unique opportunity. Athletes have a strong voice and platform; their passion for their sport often extends to a passion for the environment it depends upon. By embracing sustainable practices and advocating for responsible environmental management, they can inspire others to do the same, helping to create a more balanced relationship between extreme sports and nature.

The future of extreme sports hinges on finding a balance between thrilling adventure and responsible environmental stewardship. It is no longer sufficient to simply enjoy the thrill; athletes, organizers, and manufacturers must actively work towards minimizing their impact. Collaboration is crucial: partnerships between sporting organizations, environmental groups, and governmental bodies are key to developing comprehensive strategies for sustainable practice.

This requires a shift in mindset, moving away from a purely performance-focused approach to one that integrates environmental concerns into every aspect of the sport. From the design and production of equipment to the selection of competition venues and the promotion of responsible travel, every decision should consider its environmental implications. Ultimately, the goal is to ensure that future generations can enjoy the thrill of extreme sports without compromising the integrity of the natural world that makes them possible.

Embracing innovative solutions such as carbon offsetting programs for travel and waste management schemes for equipment can lessen the impact. Supporting and investing in research into more sustainable materials and technologies is also essential. The challenge is to demonstrate that pursuing extreme experiences doesn't require compromising environmental responsibility. The long-term sustainability of these sports depends on it. This requires a collective effort, a collaborative approach to problem-solving. If we fail to consider the environmental effects now, the future of extreme sports may be jeopardized.

The conversation around sustainability is no longer optional; it is necessary for the survival and growth of the industry. It is not about stifling the passion for extreme sports but about channeling that passion into a force for positive environmental change. The future of extreme sports isn't merely about pushing physical limits but also about pushing the boundaries of environmental responsibility, ensuring these incredible pursuits remain thrilling and accessible for generations to come. The athletes themselves hold a significant role in this change. Their voices and actions can inspire others and change the whole industry. They are not merely participants; they are stewards of the environment. Their commitment to sustainability is as vital as their commitment to performance. The combination of high-adrenaline pursuits with strong environmental consciousness is what defines a truly sustainable future.

6.5 Community Building

Extreme sports, by their very nature, foster a unique sense of community. Shared risk, intense challenges overcome together, and the exhilaration of pushing limits create powerful bonds that extend far beyond the competition arena. This isn't just about athletes; it encompasses coaches, support staff, sponsors, and even spectators who become deeply invested in the journeys and triumphs of their chosen athletes. Think of the devoted following a champion snowboarder cultivates, the shared knowledge and experience exchanged among climbing enthusiasts, or the collaborative spirit among filmmakers who document these high-stakes endeavors.

The feeling of camaraderie is palpable, forged in the crucible of intense training and shared adversity. The collective pursuit of excellence breeds a powerful sense of belonging, where individuals find strength and inspiration in the shared experience. This is particularly vital for athletes; the intense pressure, the constant risk of injury, and the relentless pursuit of improvement can be isolating. A strong community acts as a powerful buffer against this isolation, providing emotional support, encouragement, and a sense of perspective. Imagine the supportive network of fellow surfers who rally around a colleague facing a difficult recovery from a wipeout. Or the shared knowledge and support a team of mountain bikers exchange after a challenging trail ride.

Moreover, this community extends beyond the athletes themselves. Sponsors often invest in teams, not just for the marketing potential, but also for the community engagement opportunities. These investments often support training programs, community outreach initiatives, and environmental sustainability projects. Furthermore, the global nature of many extreme sports creates a unique opportunity for international collaboration and cultural exchange. Athletes from diverse backgrounds come together, sharing not only their skills and experiences, but also their perspectives and cultural heritage. Consider the vibrant exchange of ideas, techniques, and even friendship fostered during international surfing competitions. But fostering this vital sense of community requires conscious effort. Organizations and athletes need to actively cultivate inclusive environments, welcoming newcomers and fostering a sense of shared purpose. This may involve implementing mentorship programs, organizing community events, or simply creating opportunities for athletes to connect and support one another. Think of the positive impact a mentorship program can have on a young, aspiring athlete seeking guidance from an experienced professional. Or, the unifying effect of a post-competition celebration bringing together athletes, coaches, and supporters to celebrate achievements and commiserate on setbacks.

Crucially, the digital age provides new avenues for community building. Social media platforms, online forums, and streaming services offer powerful tools for connecting people from around the world who share a passion for extreme sports. However, these tools need to be employed thoughtfully. Active moderation is key to preventing negativity and fostering a positive, supportive online environment. A strong online community can offer immense support to individuals, connecting them to people who understand their unique challenges and triumphs, sharing expert advice, and providing a source of motivation. The potential for knowledge exchange and emotional support within these communities is enormous. Finally, the community surrounding extreme sports extends beyond the immediate participants. Many extreme sports venues serve as focal points for tourism, boosting the local economy and promoting environmental awareness. Sustainable practices within these communities, such as responsible waste management, support for local businesses, and advocacy for environmental conservation, are essential for long-term sustainability and community cohesion. This interconnectedness exemplifies the profound influence of extreme sports—an influence that stretches from individual athletes to the wider world. This community, nurtured and protected, will continue to inspire creativity and define generations to come.

7 Chapter 7: The Future of Extreme

Predicting the future of extreme sports is like trying to catch a wave—you can anticipate the general direction, but the specifics remain elusive. New disciplines will undoubtedly emerge, driven by human ingenuity and the persistent desire to conquer new frontiers. Imagine sports incorporating virtual reality, creating hybrid experiences where physical prowess blends seamlessly with digital landscapes. Athletes could compete in simulated environments, pushing their limits in ways previously unimaginable, with performance data analyzed and visualized in real-time, offering unprecedented insights into human capabilities. This convergence of the physical and digital will create a new breed of extreme athlete, one adept at both the tangible and the virtual.

Already, we see hints of this future in the increasing use of drones for filming and aerial perspectives, allowing for a richer, more immersive viewing experience. This technology not only enhances the spectator experience, but also assists athletes in analyzing their performance, identifying areas for improvement, and ultimately pushing their boundaries further. Beyond drones, advancements in materials science continue to revolutionize equipment design. Lighter, stronger, and more durable materials will enable athletes to perform

feats previously deemed impossible, further blurring the lines between human capability and technological assistance.
Consider the potential impact of bio-engineered materials. Imagine protective gear infused with self-healing properties, minimizing the risk of injury and extending the lifespan of equipment. Imagine specialized fabrics that enhance performance by regulating body temperature or providing targeted muscle support. These advancements would significantly impact the safety and accessibility of extreme sports, encouraging wider participation and opening up opportunities for individuals who might otherwise be excluded.
However, this technological progress isn't without its challenges. The ethical implications of advanced technology in sports must be carefully considered. Concerns about fairness, accessibility, and the potential for cheating must be addressed proactively. Maintaining the spirit of competition while embracing innovation will require thoughtful regulation and a robust ethical framework. We need to ensure that these advancements benefit all athletes, promoting inclusivity and accessibility, rather than creating further inequalities.
Furthermore, the environmental impact of extreme sports must remain a central consideration as the sports evolve. Sustainable practices in equipment manufacturing and event organization are not just good for the planet; they are essential for the long-term viability of the industry. Moving away from resource-intensive materials and embracing eco-friendly alternatives is crucial for ensuring the sustainability of extreme sports into the future. This commitment to sustainability isn't merely a matter of environmental responsibility; it's a strategic necessity for the survival and growth of the sport itself. The future of extreme sports hinges on a balanced approach, carefully navigating the thrilling possibilities of technological advancement while upholding ethical principles and environmental stewardship.
The future of extreme sports is not simply about pushing phys-

ical limits; it's about pushing the boundaries of human creativity and innovation. New sports are likely to emerge that blend elements of existing disciplines, creating hybrid forms that challenge athletes in novel ways. Imagine a sport combining elements of mountain biking, rock climbing, and free diving, requiring athletes to possess a unique blend of skills and endurance. The possibilities are endless, limited only by our imagination and willingness to explore uncharted territories. This evolution will require a fundamental shift in training methodologies, fostering multi-disciplinary skill development and emphasizing adaptability and resourcefulness. Athletes will need to be adaptable and versatile, possessing a skill set that extends beyond their primary discipline.

The continued growth of extreme sports also depends on fostering a sense of community and inclusivity. Breaking down barriers to entry and ensuring that extreme sports are accessible to people of all backgrounds and abilities is crucial for the sport's long-term health. This will require a concerted effort to promote diversity and inclusion within the industry, ensuring that participation isn't limited by socioeconomic status, gender, race, or disability. Accessibility features in equipment design, inclusive coaching programs, and targeted outreach initiatives will be critical in ensuring a diverse and vibrant future for extreme sports. This commitment to inclusivity is not just a moral imperative, it is also a strategic advantage, fostering innovation and creativity by bringing together individuals with diverse perspectives and experiences.

In conclusion, the future of extreme sports is bright, but its trajectory depends on our ability to embrace innovation responsibly, maintain a commitment to sustainability, and champion inclusivity. Technological advancements promise to reshape the landscape, but ethical considerations and environmental responsibility must guide the way. By fostering a culture of creativity, collaboration, and inclusivity, we can ensure that the future of extreme sports continues to inspire, challenge, and captivate generations to come. The next wave of extreme sports is coming, and it promises to be unlike anything we have ever seen before. The potential for innovation is unbounded, and the human spirit's desire for pushing limits ensures that this dynamic landscape will continue to evolve and surprise us all.

7.1 Emerging Sports

Drone racing has exploded in popularity, transforming from a niche hobby to a spectator sport with global appeal. Teams now meticulously design and build their drones, pushing the limits of aerodynamics and engineering to achieve mind-blowing speeds and intricate maneuvers. The races themselves are visually stunning, a blur of color and motion that captivates audiences worldwide. Imagine watching tiny, agile drones weaving through intricate courses at breakneck speeds, their pilots relying on split-second reflexes and advanced technology to stay ahead of the pack. This isn't just about piloting; it's about strategy, teamwork, and the constant pursuit of technological innovation. The competitive element creates a vibrant community of enthusiasts who continuously push the boundaries of what's possible, fostering a remarkable spirit of collaboration and ingenuity.

The rise of parkour and freerunning is another compelling example of emerging sports. What began as a form of urban exploration has evolved into a highly skilled and physically demanding discipline. Practitioners, known as traceurs, navigate urban environments with incredible agility and precision, transforming obstacles into opportunities for fluid movement. They're not just athletes; they're artists, choreographing their movements to create visually stunning displays of athleticism. The sport's appeal lies in its accessibility – anyone can participate, regardless of their athletic background – and its emphasis on creativity and self-expression. The increasing recognition of parkour and freerunning as legitimate sports highlights the shifting landscape of athletic competition and the growing appreciation for human potential in unconventional settings. This growth presents significant opportunities for the development of safety standards, coaching structures, and ultimately, a broader platform for these incredible athletes to showcase their skills.

Then there's climbing, specifically bouldering, which has transitioned beyond its niche origins to become a mainstream activity. Bouldering, a form of rock climbing without ropes or harnesses, challenges climbers to solve complex physical and mental puzzles, testing their strength, problem-solving skills and tenacity. The creation of indoor bouldering gyms has made the sport accessible to a wider audience, fostering a sense of community and driving innovation in training techniques and climbing shoe technology. The increasingly popular competitions further cement bouldering's position within the broader spectrum of extreme sports, attracting both dedicated athletes and casual participants. It underscores the growing appeal of sports which emphasize mental resilience, strategic thinking, and the pursuit of personal challenges above pure physical prowess, pushing the boundaries of what constitutes a competitive sport.

Consider the dramatic growth of e-sports, showcasing the fusion of virtual and physical athleticism. Competitive video gaming, once a solitary pursuit, has transformed into a globally recognized spectator sport, drawing millions of viewers and generating immense financial investment. Professional gamers train rigorously, honing their reflexes, strategies, and teamwork skills. These professionals dedicate countless hours to practice and analysis, showcasing exceptional cognitive abilities and strategic thinking that mirror traditional sports. This blurring of lines between the digital and physical realms highlights the evolving definition of athleticism and the innovative ways in which competition can manifest itself in the 21st century. The growth of e-sports also raises compelling questions about the nature of competition itself, highlighting the importance of mental fortitude and adaptability in a rapidly evolving digital landscape. This is a fast-paced and dynamically growing area demanding dedicated study and analysis of its societal impact.

Finally, let's consider the exciting potential of augmented reality (AR) and virtual reality (VR) sports. These technologies are poised to revolutionize the way we experience and participate in sports, creating entirely new forms of competition and athletic pursuits. Imagine immersive VR environments that simulate extreme sports, allowing athletes to train and compete in virtual settings without the inherent risks of the real world. Or picture AR overlays that enhance our perception of real-world sports, providing real-time data and feedback to athletes and spectators alike. The possibilities are truly endless, promising a future where technology and athleticism intertwine in unprecedented ways, shaping the very definition of sport and pushing the boundaries of human performance. The rapid development and refinement of this technology ensures a future where physical and digital worlds converge, potentially ushering in a new era of athletic competition that's both exhilarating and transformative.

7.2 Technological Frontiers

The relentless pursuit of speed and height demands constant innovation. Consider the evolution of snowboarding itself; from rudimentary planks of wood to the meticulously engineered high-performance boards we see today. Carbon fiber, advanced shaping techniques, and specialized bindings have all dramatically altered the sport, pushing the boundaries of what's possible and creating a cascade of related advancements in materials science and manufacturing processes. This constant refinement isn't just about going faster; it's about increasing safety and allowing athletes to perform more complex maneuvers with greater precision and control. Improved materials have contributed to better shock absorption, minimizing injury risks, a crucial element in a sport where falls are practically inevitable.

This drive for technological enhancement extends beyond the equipment itself. Data analysis plays an increasingly crucial role, with sophisticated sensor systems tracking an athlete's speed, altitude, body position, and even heart rate during a run. This real-time data provides invaluable insights into performance, allowing athletes to fine-tune their technique and identify areas for improvement. Furthermore, this data contributes to a broader understanding of biomechanics and human performance limits, potentially leading to innovative training methods and injury prevention strategies. It's a closed-loop system where technology enhances performance, which in turn drives the demand for even more advanced technology.

Virtual reality (VR) and augmented reality (AR) technologies are transforming training regimens. VR simulations allow athletes to practice complex maneuvers in a risk-free environment, honing their skills and building confidence before attempting them in real-world conditions. AR overlays real-time data onto a real-world environment, providing athletes with immediate feedback on their form and performance. This technology is also transforming how coaches analyze and strategize, allowing for remote coaching sessions and more precise performance analysis than ever before imagined. These immersive technologies are not merely supplementary aids; they are essential tools that are fundamentally shifting how athletes prepare and train.

Beyond training, the broadcast and spectator experience is being revolutionized. High-definition cameras, drone footage, and sophisticated editing techniques allow viewers to experience the thrill of extreme sports with unparalleled clarity and intimacy. The integration of data overlays provides viewers with real-time insights into athlete performance, enriching their understanding and appreciation of the sport. This heightened level of engagement further fuels the popularity of extreme sports, creating a cyclical positive feedback loop which encourages innovation and investment in the industry. Such technologies have made extreme sports far more accessible to a global audience, breaking down geographical barriers and allowing enthusiasts to feel truly connected to their chosen events and athletes.

Looking towards the future, the possibilities are truly limitless. Imagine bio-integrated sensors that provide continuous health monitoring, predicting injuries before they occur. Imagine AI-powered coaching systems that analyze an athlete's performance in real-time, offering personalized feedback and adjusting training plans dynamically. Imagine the development of new materials that make equipment lighter, stronger, and even more responsive. These advancements aren't just speculative; researchers and engineers are actively pursuing these technologies, pushing the boundaries of what's possible in extreme sports and beyond. The technological frontier in extreme sports is a dynamic space where innovation is not only encouraged but essential for continued growth and development. The future of extreme sports will be defined by the innovative technologies that empower athletes, enhance the spectator experience, and expand the possibilities for human achievement.

www.ingramcontent.com/pod-product-compliance
Lightning Source LLC
LaVergne TN
LVHW022024200926
848481LV00007B/2527
9798349282997